John Phin

Plain directions for the construction and erection of lightning-rods

Third Edition

John Phin

Plain directions for the construction and erection of lightning-rods
Third Edition

ISBN/EAN: 9783337214739

Printed in Europe, USA, Canada, Australia, Japan

Cover: Foto ©berggeist007 / pixelio.de

More available books at **www.hansebooks.com**

PLAIN DIRECTIONS

FOR

THE CONSTRUCTION AND ERECTION

OF

LIGHTNING-RODS.

By JOHN PHIN,

EDITOR OF THE "YOUNG SCIENTIST," AND THE "AMERICAN

JOURNAL OF MICROSCOPY."

THIRD EDITION.

ENLARGED AND FULLY ILLUSTRATED.

-

NEW YORK:

THE INDUSTRIAL PUBLICATION COMPANY.

1879.

CONTENTS.

PREFACE TO THE THIRD EDITION.

During the six years that have elapsed since this work was first issued, several important discussions have been held in regard to the subjects upon which it treats. The author has carefully followed all that has been written in connection with lightning rods, but he finds no reason to alter any of his original statements and directions. Several patents have also been taken out during the last few years, but none of them present any improvements of importance, the old and well known devices being sufficient to meet all demands. So that it still remains true that perfect protection may be obtained without infringing upon any patent.

The extensive demand which has existed for this book is not the only compliment which has been paid to it. The author of a somewhat pretentious volume on "Lightning Protection," has, with unblushing effrontery, appropriated several of our original illustrations, and has used his scissors freely on the text, and all without acknowledgment. This, however, was to have been expected in a book written solely for the purpose of advertising a worthless patent. *The present volume still remains, as when first issued, the only book published on lightning rods, which has not been written for the purpose of puffing some particular patent.*

The author lays no claim to novelty for any of the principles laid down in this little treatise. They are all old and well established. Many of the illustrations and methods of experimental proof are, however, original, and it is hoped that they will tend to render the subject clear and easily understood, so that, as we have always claimed, by its aid any intelligent mechanic can put up a lightning rod which will not infringe any patent, and which will afford perfect protection.

New York, May, 1879.

PREFACE TO THE SECOND EDITION.

The favor with which the first edition of this little work has been received induces the author to make a few additions of a thoroughly practical nature. These additions are so important in themselves, that he has deemed it best to throw them into the form of an appendix, rather than to incorporate them with the body of the work.

The importance of thorough protection from the effects of lightning is very generally greatly under-estimated, except perhaps in those cases where the fears of property-owners have been excited by the venders of rods, whose chief excellence lies in the amount of money which their sale brings to the manufacturers. Having carefully watched the records of the destruction of property by lightning during several years, we feel quite certain that a good rod is quite as important as a fire-insurance policy.

While giving expression to a well-founded contempt for the ordinary venders of lightning-rods, we would by no means be understood as condemning all those who are engaged in the business of manufacturing and erecting these valuable adjuncts to our houses. We are personally acquainted with dealers of thorough honesty and excellent judgment; and although we may differ from them in opinion in regard to certain points, such as the matter of insulation, terminal points, &c., yet, since we are in perfect accord as to the necessity for a substantial rod and a good ground connection, we feel that those who employ these men will not be disappointed.

At the same time, we would reiterate the statement which has been made in the body of the work, to the effect that amongst the scientific men of the present day there is no difference of opinion in regard to the general principles which should govern us in the construction and erection of lightning-rods. It is true that in a recent utterance, Prof. Henry, of Washington, whose scientific reputation stands very high, has gone back at least a century in his views. Fortunately, however, since some of his most important statements contradict the experience of every tyro in electrical experimentation, his opinions upon this subject cannot carry much weight. But with this single exception, we believe that scientific men are in perfect accord in regard to everything connected with lightning-rods.

In this little book we have endeavored to be sound and reliable, and we have not aimed at novelties, though perhaps a few new methods of illustration and of proof, relating to general principles, and of arrangement as regards the rod itself, have been introduced. In a word, we have held steadily in view the diffusion of sound practical information, and not the manufacture of a reputation for brilliancy and ingenuity.

PREFACE.

THERE are few subjects concerning which less is known by the public at large than the principles that should govern the construction of Lightning-Rods. Nor does this apply to the mere public alone. The authors of our popular scientific books, and the editors of many of those papers that attempt to give scientific instruction, are equally at fault; their directions being anything but accurate. This assertion may perhaps be characterized as "vague" and "sweeping;" but unfortunately it can, if necessary, be easily sustained by numerous pointed illustrations. And yet the question of protection from lightning is confessedly a very important one, and has engaged the attention of some of the most enlightened governments on earth. On several different occasions has a committee of the French Academy examined the subject at the instance of the government. Neither have the British left this question unexamined, as their piles of Blue-Books testify. The propriety of making such earnest investigations is obvious from the following record: In 1864, Boudin presented to the French Academy of Sciences a table showing that in France, during the period of 29 years, 2,238 persons were killed by lightning, and 6,714 struck without being killed. It is thus seen, that, during each of these years, 77 persons lost their lives, and 232 were injured. Among those struck by lightning, between the years 1835 and 1852, 67 per cent. were males, 10 per cent. females, and 23 per cent. unknown. From 1854 to 1863, 26.7 per cent. of females, and 73.3 per cent. of males, were struck.

At sea the damage done to shipping has been equally marked. A writer in *Nicholson's Journal of the Progress of Science* calculates the loss by lightning, in Great Britain alone, at £50,000 sterling annually, and Sir William Snow Harris believes that this estimate is under the mark. During the five years between 1810 and 1815, the British Navy lost by lightning no less than forty sail of the line, twenty frigates, and twelve sloops and corvettes. Indeed, it has been proved beyond all controversy that, in Europe at least, the danger of being killed by lightning is far greater than the danger of being killed by a railroad accident; or, in other words, that out of the entire population the percentage of those killed by lightning is greater than the percentage of the passengers who are killed on railroads.

And yet, in the face of this melancholy record, it remains an undoubted truth that life and property may, with certainty, be protected from injury by lightning. There are many instances where buildings furnished with rods have been struck by lightning and injured; but, so far as our experience extends, the rods have been defective in every case. In the British Navy, where the very perfect system of protection devised by Harris has been adopted, injury by lightning has become a thing almost unknown. Harris, in his work on Frictional Electricity, tells us that, in the British Navy, damage by lightning has happily ceased, but that in the Merchant Navy, where no adequate means are taken to check it, it unfortunately still continues. We regard his testimony upon this point as worthy of acceptance.

In view of these facts, it is obviously of the utmost importance that the public generally should be fully informed in regard to the principles upon which lightning-rods ought to be erected. It unfortunately happens that the information given in regard to this subject by most of our text-books on Natural Philosophy is exceedingly meagre, and as for special treatises—there is not a single one, now before the public, in the English language. It is true that we have a few pamphlets and one small book; but they can hardly be dignified with the name of treatises on the subject, since they are all written in the interest of some particular patent, and were never intended to give such information as would enable an ordinarily intelligent mechanic to erect a lightning-rod for himself.

In this country, our people appear to have derived most of their information from a class of wandering mechanics known as "Lightning-Rod Men." The object of these men, we are sorry to say, seems in many cases to be to make money rather than to afford protection from lightning; and hence, in travelling through the country, we often find a vast amount of labor and expense wasted on lightning-rods—in many cases to their positive detriment. On several occasions we have seen a lightning-rod carried alongside the metal staff which supported the weather-vane, while at the same time it was carefully insulated from it; and it is within our personal knowledge, that, on three of the new colleges devoted to the special culture of science, the lightning-rods are carefully and expensively insulated from the building by means of glass insulators. The Trustees of these institutions seem to be unaware of the fact that it is the opinion of our ablest electricians that the best way to carry a lightning-rod to the ground is to carry it down the inside of the house. The rods to which we have alluded are a standing satire on the science taught within the walls which they have been erected to protect.

Almost all the lightning-rods which are sold by these itinerant vendors are patented; and it may therefore be worth while to remind our readers that *all the essential requisites for perfect protection may be embodied in a rod which does not infringe any patent.* The issue of a patent for any device is no evidence whatever of the value of that device. Nine-tenths of all the inventions protected by patents are perfectly worthless. Every sensible mechanic knows this. Many patents are obtained simply for the sake of the opportunity which it gives the patentee to talk and argue about his invention. Indeed, a great many of these men feel a good deal as did a certain applicant for a patent, who, upon being asked by the examiner, "Of what possible use can such a device be?" replied, "Wall, I guess it is good for me to sell rights to it." There is no doubt but that all the various devices of hollow rods, twisted rods, hacked rods, and the endless variety of plans devised for attaching rods to buildings, are perfectly childish. If you have a good solid rod put up with the precautions which we detail, you need not be afraid. You are as safe as any lightning-rod man can make you. That owners of houses may in most cases find it desirable to employ competent mechanics to put up lightning-rods, we do not mean to deny. But we feel certain that a slight knowledge of the subject would have saved our house-owners, and especially our hard-working farmers, thousands of dollars out of which they have been swindled by "Lightning-Rod Men."

NEW YORK, JUNE, 1871.

LIGHTNING-RODS,

AND

HOW TO CONSTRUCT THEM.

———•◉•———

An American Invention.—Whatever differences of opinion may exist in regard to the authorship of the discovery of the identity of lightning and electricity, there can be none in regard to the fact that Franklin, and Franklin alone, invented the lightning-rod; and, strange to say, when it left his hands it was nearly as perfect as it is to-day. Notwithstanding the many patents that have been issued for alleged improvements in lightning-rods, we would as soon trust our property to a rod constructed after the directions given by the inventor as to one made according to the latest devices, covered by the broad seal of the United States Patent-Office. It is true that during the last half-century we have learned something about rods and their mode of action, and have acquired a more perfect knowledge of the special points which demand attention, in the effort to secure perfect safety. But it will be found, on examination, that Franklin's old rod embodies all the points necessary for a perfect conductor, and that, consequently, we are not at the mercy of any patent-right monopoly in this matter.

Are Lightning-Rods Really a Protection?—There are many instances on record where buildings protected by rods have been struck and injured. But this is not to be wondered at, when we reflect that fully one-half—nay, perhaps three-fourths—of all the rods now actually erected, violate the fundamental principles upon which their efficiency depends. Besides serious errors in regard to arrangement and continuity, it will in general be found that it is only by the merest accident that a good ground connection is ever secured. This point will be more fully discussed in a subsequent paragraph. Meanwhile, the following facts prove irrefragably the great value of well-constructed rods. The Cathedral of St. Peter, in Geneva, although so elevated as to be above all other buildings in the neighborhood, has for three centuries enjoyed perfect immunity from damage by lightning; while the tower of St. Gervaise, although much lower, has been frequently struck. This doubtless arises from the fact, that all the towers of St. Peter are accidentally furnished with very perfect conductors. The great column of London known as the Monument, erected in 1677 in commemoration of the great fire, although over two hundred feet in height, has never been struck; while much lower buildings in the vicinity have not escaped. The Monument, however, is protected by a most perfect conductor; the upper end terminating in a vase from which proceed numerous metal plates designed to imitate the appearance of tongues of flame. The vase communicates by means of stout bars of iron with the metal staircase which descends through the middle of the column, and terminates in the ground. A still more striking instance of the value of lightning-rods is a church on the estate of Count Orsini, in Carinthia. This building was placed upon an eminence, and had been so often struck by lightning that it was deemed no longer safe to celebrate divine service within its walls. In 1730, a single stroke of lightning destroyed the entire steeple; after it had been rebuilt, it was struck on an average four or five times a year, without counting extraordinary storms, during which it was struck from five to ten times in a single day. In 1778, the building was reconstructed and furnished with a conductor; and, according to Lichtenberg, up to 1783,—that is to say, during the space of five years,—the steeple had been struck only once, and this stroke had fallen upon the metallic point without producing any damage. For two or three years after its erection, the church of St. Michael in Charlestown had been frequently damaged by lightning; a conductor was attached to it, and during the following fourteen years it had not been injured. The steeple of St. Mark's in Venice has a height of 340 feet, and was frequently struck by lightning until a proper lightning-rod was applied to it, since which time it has not been injured. These facts leave no room for doubt in regard to the great value of lightning-rods.

What the Lightning-Rod Should do.—It is not our purpose to enter into an attempted exposition of the theories of electricity, its phenomena and laws. For these we must refer our readers to any of the ordinary works on Natural Philosophy. At present we propose to deal with bare facts, and to give concise directions for practical guidance. The function or office of the lightning-rod is twofold. In the first place, it acts as a means whereby the accumulated electricity existing in the atmosphere is silently drawn off, and allowed to pass into the earth, and thus prevent an explosion; and, in the second, it acts as a path by which explosions, lightning flashes, or disruptive discharges (as they are more properly called) may find their way to the earth freely, and thus be carried off without any danger of their acting with mechanical violence, as they are certain to do when made to pass through what are called non-conductors. Experience teaches us that, so long as a discharge of electricity passes off through a wire that is large enough to carry it safely, it does not cause any damage or give rise to the exhibition of mechanical violence. A spark from the prime conductor of an electrical machine, if passed through a moderately fine wire, does not injure it; if passed through a thick card, it will pierce it; and, if passed through a small block of wood, it will rend it asunder. On the occasion of every thunderstorm, there is a large quantity of electricity to be conveyed from the clouds to the earth, through the air, which is, in general, a very poor conductor. This electricity always tends to pass by the easiest path, or, as electricians say, *the line of least resistance*. The resistance of any line may be lessened by various circumstances, such as the presence of hot vapors, as from chimneys, heated haystacks in the open field, or heated haymows in the barn; the existence of a line of carbonaceous matter, such as exists in a column of smoke; the presence of a tree with its leaves and sap, or of a house with its chimneys; or the fact that the air has been rendered moist by the passage of a shower of rain. So difficult is it, however, to detect the circumstances which render any particular path more easy than others, for the electricity to follow, that we are often unable to give a reason for its following a particular course, and the action of this mighty force seems to us like a mere freak. Such ideas are, however, entirely wrong; and we may accept most implicitly the statement, that the flash will always take the easiest path, and it must be our duty to determine beforehand what this path shall be, and to make it so easy and so perfect that the resistance will not cause the electricity to produce the slightest mechanical violence. And if we do so, then, just as one may hold in the hand a slender wire while a powerful charge of electricity is sent through it,—even so we might hold in our hands a good lightning-rod at the instant of its being struck, and yet receive no injury.

The Proper Material for Lightning-Rods.—The foregoing considerations clearly show that lightning-rods should be made of the very best conducting material, and of a size sufficient to carry off the heaviest discharge that is ever likely to fall upon them. Of all well-known substances, metals are the best conductors; but even amongst metals there are great differences in this respect; some metals, according to Becquerel, conducting nearly 75 times better than some others. But, on examination, we find not only great differences in the conducting powers of the metals, but great differences in the estimates of different observers. Thus it will be seen that Reiss estimates the conducting power of silver at four times the value which is given to it by Ohm; both estimates being referred to copper as a standard. It must be noted, however, that there is a certain marked uniformity of result in the determinations of all these savans. They all place gold, silver, and copper very high in the scale; while the values assigned to platinum, iron, tin, and lead are very low. We may perhaps note that the determinations made by Lenz are, in general, regarded as being amongst the most reliable. There are, however, but two metals whose claims are worth considering,—copper and iron. The others are too costly.

The following table gives the relative conducting powers of the different metals according to the best authorities:—

Table Giving the Conducting Power of Metals for Electricity.

METAL.	NAME OF OBSERVER.					
	BEC-QUEREL.	OHM.	DAVY.	LENZ.	REISS.	POUILLET.
Copper,	100·00	100·00	100·00	100·00	100·00	100·00
Silver,	73·60	35·60	109·10	136·25	148·74	
Gold,	93·60	57·40	72·70	79·79	88·87	103·05
Zinc,	28·50	33·30				
Platinum,	16·40	17·10	18·20	14·16	15·52	22·50
Iron,	15·80	17·40	14·60	17·74	17·66	{ 15·60 18·20
Tin,	15·50	16·80		30·84	14·70	
Lead,	8·30	9·70	69·10	14·62	10·32	
Mercury,	3·45					2·60
Potassium,	1·33					
Brass,		28·00		29·33	27·70	{ 15·20 23·40
Palladium,			16·40		18·18	
Cadmium,					38·35	
Nickel,					13·15	
Cast Steel,						{ 13·00 20·50

Comparing the estimate of the different authorities given, we find the estimated relative values of copper and iron to vary between 14.6 and 18.2 for iron to 100 for copper.

It is found, however, that different samples of both copper and iron vary very much in their conducting powers. Pure copper being taken at 100, it will be found that the conducting power of even the best selected samples of commercial copper is less than this by from five to 15 per cent.; while the ordinary commercial article falls below this standard by from 30 to 60 per cent. On the whole, we are perhaps safe in taking the relative conducting powers of iron and copper as one to four. That is to say, that, if a rod of copper of a given size is found to be sufficient, then an iron rod, to be equally safe, ought to weigh four times as much per lineal foot.

Proper Size.—Great differences of opinion exist in regard to the size of rod necessary to insure safety. The old directions by the French Academy of Sciences, named a rod of from one-half inch to one inch square as a safe conductor. The following extract from the last report gives their latest views upon this point: "A discharge of our electric batteries is capable of melting several yards of very fine iron wire. A flash of lightning is capable of volatilizing more than one hundred yards of bell-wire, or of the wire that is usually employed in connection with the hammers of public clocks. In 1827, upon the packet-boat *New York*, a surveyor's chain forty yards long, made of iron wire a quarter of an inch in diameter which served as a lightning-rod on the vessel, was melted by a flash of lightning and dispersed in red-hot fragments. No instance has ever occurred in which lightning has raised to a red heat a bar of iron some yards long, and four-tenths of an inch square, or having a section of one-sixth of a square inch. Hence a square rod of iron, the side being four-tenths of an inch, has been adopted in the construction of lightning-rods." This shows that the enormous conductors recommended by Professor Henry and others are unnecessarily expensive and clumsy.

The sizes usually adopted have not been determined by very careful investigation, and cannot be regarded as safe guides in practice. Ordinary rods, made of iron, are about half an inch square; and few copper ones are more than equal to a wire a quarter of an inch in diameter. We are satisfied, however, that it is not in size that ordinary rods are deficient, but in continuity, arrangement, and ground connection. It is of the utmost importance that lightning-rods should be equal to any demand that may be made upon them; and, after carefully examining the subject, we feel satisfied that the conclusions reached by the French Academy are very nearly correct. Instead of a square rod, however, we would recommend a strip or bar one inch wide and three-sixteenths of an inch thick, or, if it were a quarter of an inch thick, it would do no harm. A rod of the first-named dimensions will weigh ten ounces per lineal foot. The weight of the second will be thirteen ounces per foot. If a copper rod be used instead of an iron one, it may be made much thinner; but we could hardly recommend its reduction to the fifth or sixth of the size of the iron rod, as theory would indicate. A copper rod, having the width of one inch, should be not less than the twelfth or the tenth of an inch thick. A rod of the latter size weighs six ounces (.384 lbs.) per lineal foot.

The Best Form for the Rod.—More nonsense has been written in regard to this subject than in regard to any other connected with lightning-rods, except perhaps that of insulation. It has been the subject of several patents, and has furnished an immense amount of capital to itinerant venders of rods—so much so, that we deem it necessary to present clearly the reasons which lead us to believe that the efficiency of a rod depends, other things being equal, solely upon the amount of metal which it contains, and not at all upon the form that may be given to it.

If we examine the rods ordinarily found in market, and puffed by those who have invented them, we shall find that, instead of being solid bars of a square, round, or merely flattened form, they are tubes, twisted ribbons, or bars whose cross-section has the form of a star. And if we ask why these complicated and expensive forms have been adopted, we shall be told that it is for the purpose of obtaining the greatest amount of surface with the least amount of metal, and this is done because electricity always resides on the surface. Those who reason in this way, however, prove clearly that they have never studied the subject, else they would be aware of the fact that while *static* electricity, as it is called—that is, electricity at rest—always disposes itself on the surface of bodies charged with it, electricity in motion pervades the entire substance of the bar through which it passes, and consequently the power of such a bar to convey electricity is measured by the quantity of metal that it contains, and not by the extent of surface that it presents. Pouillet showed this in a very clear and decisive manner. He measured the conducting power of a fine wire of cylindrical form—the form that presents the least possible surface in proportion to its cubic contents—and then, having flattened and annealed it, he tested it again. Its surface was enormously increased, but its power to conduct electricity was lessened rather than otherwise; this diminution being probably due to the fact that the wire was increased in length, and, consequently, its cross-section was somewhat diminished. An experiment, equally decisive, and perhaps somewhat more easily performed, is at the command of every one who has access to a small electrical machine and a two-quart Leyden jar. Take a fine gold wire, say the one-fiftieth of an inch in diameter. This wire will present nearly the same surface as a ribbon of metal 1-32nd of an inch wide. The wire will

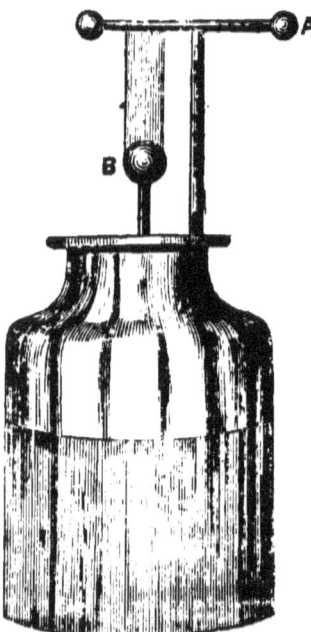

carry off, without being injured, any charge that can be imparted to the jar. If, however, we pass the charge through a strip of gold leaf having several times the surface of the wire, it will be completely burned up and dissipated.

The following very interesting experiment shows clearly the difference between the action of electricity at rest and electricity in motion, and exhibits also the fact that surface is of no avail in carrying off a charge of electricity. Take a Leyden jar of large size—say two gallons—having the usual knob and other arrangements, shown in Figure 1. In the wooden cover insert a glass tube carrying at its upper end a wire, lying horizontally across it, this wire having a ball at each end, so that the discharge may take the form of a spark or an explosion and not pass off silently. Between the horizontal wire and the knob of the jar stretch a strip of gold-leaf (B), and

FIG. 1.

charge the jar in the usual manner. So long as no discharge passes *through* the gold-leaf, it will remain uninjured, although it is evidently charged as intensely as the machine can charge it. But if we discharge the jar by laying one ball of the discharger on the outer coating of the jar and the other on the knob (A), the gold-leaf will be destroyed. But a wire the thirtieth of an inch in diameter, although presenting far less surface, would have carried off the charge with impunity. In short, to end the matter, we may state that while no flash of lightning has ever been known to injure a copper rod half an inch in diameter, we have frequently, by artificial means, dissipated an amount of gold-leaf that presented a greater amount of surface.

The French Academy, although they give no specific directions as to form, always speak of a square rod. The well-known authority, Sir William Snow Harris, in his work on "The Nature of Thunderstorms," says: "Provided the quantity of metal be present, the form under which we place it is evidently of no consequence to its conducting powers, since it would be absurd to suppose that a mass of metal, under any form, did not conduct electricity in all its particles ; indeed, we know that it does so, and that it is impossible to fuse by electricity *a portion only* of a homogeneous metallic plate of uniform thickness." For ourselves, we recommend the form of a flat ribbon, from purely mechanical reasons. A bar of metal in this form is more easily bent, more conveniently attached to the building, and is less conspicuous, than when in any other form. Its conducting power is neither improved nor impaired.

Arrangement of the Rod in Regard to the Building.

—The arrangement of the rod deserves the most careful consideration, not only because upon it depends in a large measure the efficiency of the conductor, but because the amount of metal used determines largely the expense of the rod. The arrangement usually adopted consists in elevating from each chimney a tall spike, and connecting the several spikes or points by rods which either join the main conductor, or descend directly to the earth. This is very well, and is in accordance with the directions of the French Academy of Sciences, but is not quite as perfect as it might be.

It is very obvious that the most perfect, though not the most economical, arrangement would be to cover the entire building with a sheet of metal. Then it would be impossible for the lightning to strike any point without falling upon a good conductor. But such an arrangement being obviously inadmissible, we must so arrange matters that the most exposed points shall be the conductor and its connections. With this object in view, the French Academy recommended the use of rods elevated nine to twelve feet above the building ; and, after a good deal of experiment and observation, came to the conclusion that a rod was capable of protecting efficiently a space covered by a radius equal to twice the height of the rod, above the most elevated part of the building. For ourselves, a very careful examination of the conditions which

arise during a thunderstorm, leads us to place no confidence whatever
in elevated rods, and to depend wholly upon so arranging the conductor that every part of the building shall be protected.

In a former paragraph we stated that the lightning stroke always
follows the line of least resistance, and that this line may be determined by the presence of vapor, smoke, moisture, etc., forming ascend-
ing or descending streams or horizontal layers; and it is in this way
only that we can explain how it happens that a house furnished with
very elevated conductors, will be sometimes struck on the gable, or why
a mast unfurnished with a conductor, will be struck, while a taller
mast, distant but a few feet, and provided with a conductor, is not
struck, and evidently has not served to carry off the discharge.   If the
line of least resistance were always the *shortest* line mathematically,
the rule of the Academy would hold good.   But let us consider the case
of the building shown in the acccompanying engraving (Figure 2),

FIGURE 2.

which represents a barn furnished with a conductor and exposed to a
thunderstorm. We will suppose that the barn has been newly filled with
hay, which is giving off the warm vapor that is pouring out of the end
window, and forms an invisible band of conducting matter between the
cloud and the barn, marked out in the engraving by dotted lines—the
direction of the wind being shown by the arrow. In this case, the flash
will pursue the longer path between c and d, in preference to the shorter
one between a and b, and the barn may be set on fire, although furnished
with one of the best of rods. The same deflection of the bolt might
be caused by rain, by a column of smoke, or by the fact that a part of
the building had been moistened, while other parts had been kept dry,
as frequently happens when the rain is driven in particular directions
by a strong wind. In considering this subject, it must be borne in mind
that the flash does not leave the cloud and pursue a sort of experimen-

tal path through the atmosphere, leaping from point to point like a boy
picking his way through a swamp—now going forward, now sideways,
or sometimes going backward for the purpose of obtaining a new point
of departure. On the contrary, before the slightest disruptive action has
taken place, the electricity has, by the power of induction, marked out
for itself the precise path that it will take, and it pursues that path in-
flexibly. Every turn, every break, has been clearly defined ; and it would
be of no use to introduce into the neighborhood of this path any good
conductor, unless the path itself were thereby electrically shortened.
No additional elevation given to the conductor shown in the engraving,
and no spike or spikes raised from other parts of the ridge of the roof,
would have tended to render this path shorter.   But if the conductor
had been simply carried along the ridge, and down the edges of the
roof at the gable, a way of escape would have been opened, and the
barn would have been saved.

In the accompanying engraving we give a figure of a small house, in
which the arrangement of the lightning-rods is shown by dark lines.
It will be seen that all spikes and points are avoided as being costly,
unsightly, and inefficient, as will be more fully explained in a subsequent
paragraph, treating of the termination of the rod in the air. Each
chimney is surmounted by a cast-iron cap, to which the rod is attached,

FIGURE 3.

and, after descending to the ridge, the rod is led along it in every
direction, and caused to descend along the edge of the roof at each
gable. It is of course obvious that if the building be crowned with a
Mansard roof, and finished with crestings, these will answer every purpose of a lightning-rod so far as they go. But all such crestings, as

well as the waterspouts, tin or other metal roofs, eave-troughs, etc., must be brought into metallic connection, so that the building may form, electrically, a homogeneous mass.

A single good rod, carried to the ground and placed in perfect electrical contact with the earth, as hereafter directed, is as good as a dozen rods   We often see a building provided with several rods, all descending into the earth.   Such an arrangement shows great ignorance on the part of those who directed the matter.   In the first place, the extra metal employed could be used to much better advantage by being distributed over the upper portion of the building ; and, in the second, good earth connections are too expensive to be needlessly multiplied.   The heaviest item of expense in the construction of most rods is, or ought to be, the earth connection, if it be properly made.   One good one is amply sufficient ; half a dozen poor ones are of no use at all.

It is obvious that the rod should be carried from the roof to the ground by the shortest possible way.   Some have even recommended that it be carried down the chimney—a plan in which there is nothing objectionable, provided it happen to be the most convenient.

Barns.—Before leaving the subject of the arrangement of lightning-rods, it may be well to say a few words in regard to the peculiar conditions found in barns, ice-houses, and similar structures, and the best means of protecting them.   It has long been known that barns, and stacks of hay and grain, are peculiarly liable to be struck by lightning ; and the reason of this is not difficult to discover.   It rarely happens that grain or hay is perfectly dry when put away, and the peculiar character of the atmosphere that prevails during the approach of a thunderstorm furnishes just the conditions necessary to produce heating or fermentation, and the consequent production of a stream of dense, moist vapor, having comparatively high conducting power.   This stream of vapor forms in reality a conductor leading from the clouds to the barn, or stack, but *ending there*—the very condition that we should seek to avoid ; for while a conductor that leads *past* any object is a sure protector, a conductor that leads *to* an object, and ends in that object, is an equally certain destroyer.   Our aim must, therefore, be to cause any natural conductor that may exist, to coincide with our artificial conductor, and unite with it.   This is best accomplished by furnishing all barns designed for the storing of hay with a ventilator on the ridge, and leading the lightning-rod to this ventilator.   Where there are no ventilators but doors, or large open windows in the gable, as in the barn shown in Figure 2, the most efficient method will be to carry a stout rod horizontally from the ridge, for a distance of two or three feet. To carry it vertically in such circumstances will be of no use, as it will then be led *out* of the current of conducting vapor, while if it project horizontally, it will be carried into that stream, whether the vapor ascends vertically, or is carried off in the manner shown in our sketch.

Where ventilators are placed on a barn for the purpose just mentioned, they should take the form of an observatory, or cupola, and the rod should be carried down at least four sides or corners of this cupola.

Should Lightning-Rods be Painted?

—It is not many years since one of our popular journals, whose main claim to public favor is based upon its supposed reliability in scientific matters, gravely informed its readers that lightning-rods should not be painted because the electricity passed along the surface, and to paint the rod would be to render the surface a non-conductor and of course to destroy the efficiency of the rod! And in a circular published by the well-known E. Meriam, of Brooklyn Heights,—a circular which was not only published in our most widely circulated public journals, but was distributed by the thousand by Mr. Meriam himself,—this idea is advocated. As Mr. Meriam's name has been very generally regarded as of pretty good authority, it is no wonder that this idea has in many quarters taken strong hold of the public mind, and it may not be out of place to state the grounds upon which it is regarded as erroneous by all scientific men.

In the first place, it is an erroneous idea that it is the surface alone of a rod which has the power of conveying the electricity. This we have shown very fully when discussing the proper size and shape of the rod. In the second place, the surface of the *metal* would remain metallic even if covered with a dozen coats of paint. The fact is, that every lightning-rod is covered with a coating of non-conducting material, whether we will or not; and this material is quite as good a non-conductor as paint! We refer, of course, to the air which of necessity surrounds every rod, and which is well known to be a non-conductor. Even in the case of static electricity, which resides almost exclusively on the surface, it is found that to paint or—what is the same thing—to varnish the conducting surfaces does not diminish their power to receive, retain, or convey a charge of electricity. All electrical apparatus is carefully varnished with one of the best of non-conductors,—shellac varnish. Even the prime conductor, which it is desirable to keep in the very best condition for receiving and conveying electricity, is always thoroughly varnished ; and this operation has never been known to affect its efficiency. Submarine cables are always most carefully insulated throughout their entire length, and yet this operation does not detract from their conducting power. The truth is, that arguments innumerable might be advanced to show that paint does no harm ; but the above will suffice. In conclusion, we would merely ask the opponents of painting why they are all advocates for insulation ? It seems to us, that, if a coat of paint is sufficient to destroy the conducting power of a rod, the insulators themselves must detract from its efficiency. since it must be very difficult for the lightning to dodge around them ! !

To our readers we would say, paint your rods, no matter of what material or of what construction. By so doing, you will protect them from the influences of the atmosphere, which in the case of some metals, such as iron, is very destructive. Moreover, you can thus cause them to resemble, in color at least, the building to which they are attached ; and, as lightning-rods can never be an ornament to any edifice, you will thus be enabled to hide what would otherwise be a disfigurement. In a series of directions by Professor Henry, of the Smithsonian Institute of Washington, communicated to a gentleman of Boston, and extensively published by the papers throughout the Union, it is recommended to paint the rods with *black* paint, on the ground that such paint is a good conductor. The alleged fact is not true, and the directions are injudicious. Black paint does not conduct any better than any other paint ; this we have tested by experiment. To attach a black rod to a light-colored house would be outrageous.

Termination of the Rod in the Air.

—At an early period in the history of the lightning-rod attention was called to the rapidity with which points give off and receive electricity, and also to the fact, that when a Leyden jar is discharged by means of a sharp point, we no longer have that loud explosion that occurs when the ends of the discharger are furnished with balls. Reasoning from these facts, Franklin, with his usual sagacity, recommended the use of pointed conductors, and they have been used ever since, although at first they were most bitterly assailed by certain court favorites in England. This famous discussion, which was known as " The War of the Knobs and Points," ended in the almost universal acceptance of the form recommended by Franklin; and wherever lightning-rods were used, they were made to terminate in points.

Some persons, seeing that a single point is so efficacious, have advised that several be used, upon the ground that two points would do twice as much work as one. Those who adopt this view forget that the more we multiply the points, the more nearly do we approach to the form of a ball or blunt rod. Experiment shows very clearly that two points are *not* as efficient as a single one, and that the more points we have, the more the conductor acts like a ball.

After all, however, the term point is a comparative one. That which would be a fine point when opposed to the side of a house, becomes a blunt rod when used in connection with the ball of a Leyden jar. And how large may that point be which will bear to a thunder-cloud, extending over thousands of acres, the same relation which a fine pin bears to the knuckle of the human hand, or even to the conductor of an ordinary electrical machine ? Under such circumstances, the entire railings of a Mansard roof become. virtually a point ; and a space equal to that exposed by the top of a chimney, is certainly small.

Considerations such as these have led many of our best electricians to discard the use of points, and to advocate the adoption of blunt conductors, and even balls. Thus De La Rive, in his " Traité d'Electricité," (Tome III, page 161), speaking of the aerial termination of lightning-rods, says : " We even believe that a small sphere of gilt copper would be preferable, for it would resist more powerfully all destructive agencies, and its radius of activity would not be much less than that of a moderately sharp point. It would also have the advantage, that its influence would be exerted in all directions,—an important feature in the not very rare cases, in which the lightning strikes buildings laterally." The truth is, that the point is not one of very great importance; though, in regard to the causes which induce lightning to strike buildings laterally, those who have studied carefully the section on the arrangement of the conductor, will be able to appreciate the probable efficiency of a gilt ball in warding off such strokes.

The ablest and clearest exposition of this subject that we have met, is that given by the Committee of the French Academy, in their last report upon the protection of buildings from lightning. This committee consisted of MM. Becquerel, Babinet, Duhamel, Fizeau, Edm. Becquerel, Regnault, le Marechal Vaillant, and Pouillet,—all names of men standing in the front ranks of scientific Frenchmen. After describing the main features of a well-constructed lightning-rod, they go on to say : " Supposing a lightning-rod to be erected in accordance with these conditions, let us examine, in a general manner, the phenomena which would occur during thunderstorms. The electricity developed by influence in the subterranean water-bed, instead of accumulating there, as we have formerly described (when no conductor was present— *Trans.*), finds the foot of the conductor, and rushes to escape by this path ; for, even in the interior of a solid metallic bar, however long it may be, the electric fluid expands itself with a rapidity which may be compared to the velocity of light. It is in this way that the electric fluid in the subterranean water-bed is attracted by the cloud, and suddenly accumulates in the upper part of the lightning-rod. It there produces certain curious phenomena, which it is necessary to detail.

" If the lightning-rod ends in a very fine and sharp point of gold or platinum, the fluid attracted by the cloud exerts against the air, which is a bad conductor, a pressure so great that it will escape, producing at the same time a luminous star, visible in darkness. The rays diverging from this star diminish in brightness as they recede from the point ; they are rarely visible for a greater distance than fifteen to twenty centimetres. The air is powerfully electrified by them, and there is hardly any doubt but that these particles of air, thus charged with fluid from the point, that is with the attracted fluid, are carried even to the cloud itself, if the air be calm, and thus neutralize a portion, more or less appreciable, of the fluid with which the clouds are charged. It is this neutralization which is called the preventive action of the rod.

" At the same time that the sharp point gives rise to the luminous star (*aigrette*), the escaping electricity often attains such a degree of intensity that the point becomes heated even to fusion; in this case, gold, and even platinum itself, although much less fusible, fall in large drops along the copper or iron to which they are attached. When a lightning-rod has thus lost its sharp point, and when its upper end has become a mere large, fused button of gold or platinum, we naturally inquire if it has not been rendered useless? This question may be answered in the negative, provided the rod continues to fulfill two essential conditions, which are: 1, that it shall be continuous, and 2, that its lower extremity shall communicate largely with the subterranean water-bed. In losing its point, the lightning-rod loses only its preventive action. The star is now produced only under the influence of a much more powerful electrical action; and fusion; which depended chiefly upon the fineness and sharpness of the point, is renewel only with great difficulty, whilst, at the same time, it leaves things precisely as they were before. The air is no longer electrified by the star in a luminous form; this part of the preventive action has disappeared; the other part, that which depends upon the electrifying of the air by contact with the upper part of the rod, is probably much lessened. Finally, if, as often happens, the electrified air from the star, and from the rod itself, should be carried by the wind away from the cloud, the preventive action is often reduced to nothing, which, however, is not greatly to be regretted. The conclusion is, that in losing its sharp point, a lightning-rod, in reality, loses but a very trifling advantage.

"These considerations led the Commission of 1855 to advise that the upper end of the rod be terminated by a cylinder of copper, eight-tenths of an inch in diameter and eight to ten inches long, the summit being tapered off into the form of a cone one and a quarter or one and a half inches high. This copper cylinder is attached to the rod by means of a screw, and is brazed to it so as to make metallic contact certain.

"Taking, now, for our example, a rod, of which the summit is terminated by a cone of copper, and leaving out of consideration what is known as the preventive action, let us examine the phenomena which are produced during storms.

"The copper cone may still sometimes present the spectacle of a luminous star, but less often than sharp points of gold or platinum; even then it resists fusion by reason of its form, and, above all, by reason of its great conducting power. If the lightning should strike, it is by the copper cone that it penetrates the rod, and it is by the rod that it passes to the subterranean water-bed and is neutralized. The two points of departure of the lightning reside, one upon the cloud and the other upon the summit of the lightning-rod; there is no appearance of light or of electricity in any other part of the circuit. The current produced by the lightning passes through the entire substance of the conductor, just as the current produced by an electric or voltaic battery passes through an iron wire of sufficient diameter.

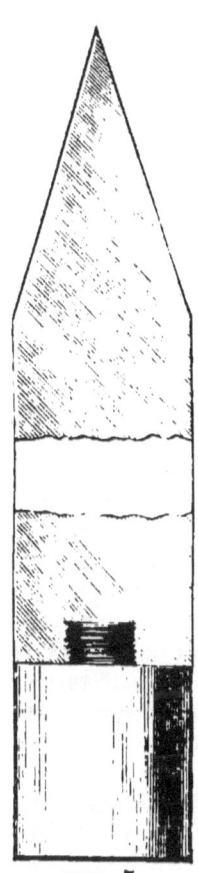

FIG. 5.

"These are the phenomena attending an ordinary stroke of lightning, only it does no damage to the rod or to the edifice which the latter protects ; it resembles the innumerable discharges which, during a thunderstorm, occur in the middle of the atmosphere.

" Figure 5 gives a section of the copper cylinder attached to a stiff iron rod, which supports it above the building."

These conclusions we consider perfectly judicious, and they show us the utter inutility of those numerous points which we find sticking up all over our best houses, and which so greatly disfigure them. In regard also to the necessity for gilding the copper cone, the popular notions are equally at fault. We have examined the subject carefully, by experiment, and have found that a common, rough, cast iron ball draws sparks almost if not quite as well as a copper ball gilt and unvarnished. Even a painted metallic ball draws a spark without any appreciable inferiority to a bright brass ball, so that, for ourselves, we do not see the necessity for either gilding or platinizing the end of the rod. If our readers desire to have their rods terminate in gilt points or cones, we have no objections to offer, except on the score of expense and unsightliness.

The terminations which we recommend are these: The rod itself, painted, for the ridge and gables, and cast-iron caps on the chimneys. Such caps may be made highly ornamental instead of being a disfigurement, as are the ordinary points with which most houses bristle, while, at the same time, they serve a most useful purpose in the preservation of the chimneys and the prevention of all loosening of the bricks—a most disagreeable, unsightly and dangerous, though a most common occurrence. All we have to say, however, is that if you desire a number of gilt points sticking up all over your house there is no objection to your putting them up. Our opinion is, that they do little or no good, though they certainly do no harm, further than to disfigure the building. Where the roof is finished off with railings, crestings, finials, &c., no better termination than these can be desired.

The Best Method of Attaching Lightning-Rods to Buildings.—It is a very prevalent opinion that lightning-rods should be carefully insulated from the buildings to which they are attached, and hence most rods are made to pass through glass tubes or insulators, the avowed object being to prevent the electricity from

passing into the building. The extreme worthlessness of such arrangements ought to be obvious to any person that ever observed a flash of lightning, and the positively dangerous character of the insulators will be apparent to every one that carefully studies the principles which govern conduction and induction. That a flash of lightning which will pass through a hundred feet of air should be unable to pass from the lightning-rod to any other object, merely because an inch or two of glass is in the way, is a proposition too absurd to find favor with any sensible man. When, moreover, we find that during a thunderstorm this insulator is in general deluged with rain, it does not require much acumen to see that the little efficiency that it ever had will, under such circumstances, be totally destroyed. We may, therefore, set it down as tolerably certain that none of the insulators in ordinary use answer the purpose for which they are intended; or, in other words, that no insulator can be invented which will prevent the lightning from leaving the rod if there is any inducement for it to do so. The latter statement will not appear extravagant when we remember that the discharge from a powerful Rhumkorf coil will pierce through five inches of solid glass, so that, if a lightning-rod were entirely cased in a glass tube, an artificial flash of lightning can be produced which will pass through it as easily as a spark from one of the old machines would pierce a card! And if such a result can with safety be produced by artificial means, in a room filled with people, what would be the limit to the effects produced by natural means where thousands of acres of cloud-surface are engaged? Insulators do very well to talk about, but as a security against Heaven's artillery they are powerless.

But not only are insulators worthless—they are positively dangerous, if the principle upon which they are adopted is fully carried out, which, however, is but rarely done. A very little consideration will show this. Thus if a house be furnished with a carefully insulated lightning-rod, and should also have any large surface of metal, such as a tin roof, an extensive system of gutters, or such like, connected with it, it is easy to see that the house must resemble a large Leyden jar, of which the tin roof, or other mass of metal, constitutes one coating, and the lightning-rod and the earth constitute the other, while the insulators and the dry material of the house represent the glass of the jar. If both the outside and inside of this jar (the tin roof and the earth) had been connected together, it would have been impossible to have brought one coating into a condition opposite to that of the other. But the rod being carefully insulated from the roof, it is obvious that the inductive action of the cloud will bring the roof and the earth into opposite conditions; and if a man were to form the path of least resistance between them, the discharge would take place through his body, and he would probably be destroyed. It is obvious, then, in the first place, that lightning-rods should be connected with all large masses of metal which may exist in or upon the house, such as metallic roofs, tin or iron gutters, or pipes, iron railings, etc. In the second place, the rod should be attached to the house in the neatest and least obtrusive method possible. If the rod be flat, it may be pierced with small holes and tacked directly to the building ; but a better way, both for round, square, and flat rods, is to employ properly shaped staples of stout wire. These staples may be driven into the studding of wooden houses or into the joints of brick walls, and, when properly painted, will not present an unsightly appearance.

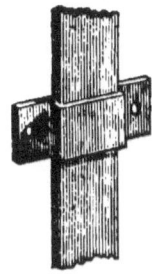

Where something better than mere staples is desired, the device shown in the annexed engraving may be employed. It consists of a strap of metal of the same kind as that of which the rod is made, bent as shown in the engraving, Figure 4, and pierced with two holes, whereby it may be attached to any structure by means of a couple of screws as shown. The advantages of this device are, that it does not weaken the rod, is not unsightly, permits the rod to slide on the building as it expands and contracts by heat and cold, and permits it to be easily applied or removed without injury to the building. Staples, however, are nearly as good and much cheaper.

FIG. 4.

Termination in the Ground.—Upon the perfection of the ground termination mainly depends the value of the lightning-rod. If this be defective, no other good features can possibly make up for it. And yet, so little is it understood, that a careful examination of a very large number of rods leads us to believe that fully one-half the lightning-rods in existence are defective in this respect, and, consequently, furnish but an insufficient protection.

In discussing the proper size of the rod, we stated that a rod of copper need not be as large as a rod of iron, for the simple reason that copper is a better conductor than iron. Now, it is well known that the terms conductors and non-conductors are only comparative ; all conductors resist the passage of electricity to a certain degree, and all non-conductors allow it to pass, though with difficulty. Iron conducts many thousands of times better than water, but water conducts many thousands of times better than dry earth. And just as we are able, by increasing its size, to make an iron rod conduct as freely as a copper one, so, by increasing the volume of water or soil employed to carry off the electricity, we can make it almost as efficient as the metallic conductor. These facts lead us to the following conclusions : 1. The end of the rod ought to be made to terminate in a layer of soil that is permanently *wet;* and 2. The end of the rod ought to expose to this soil as large a surface as possible.

Permanently moist earth is to be attained only at considerable depths, —say at the level of the water in the wells in the vicinity. Unless we reach this point, we can never be sure that our rod does not terminate in dry or but slightly moist soil ; consequently, no effort should be spared in sinking the rod to a sufficient depth. This is most easily accomplished at the time when the foundations are laid ; and we would advise all builders to sink the lightning-rod termination when they sink the foundation. A short portion of the stem may be allowed to rise above the ground, and the conductor may be arranged and attached at a subsequent period.

In advising that the termination of the rod should expose as large an extent of surface to the soil as possible, we might, at first sight, seem to be departing from the principles laid down when we discussed the proper form of the rod ; but the necessity for exposing a large surface of rod to the soil to which the electricity is to pass will be obvious when we consider that this surface is the measure of the *section of solid soil employed to carry off the electricity.* Surface, in this part of the rod, performs a function very different from anything performed by that part of the surface of the rod that is in contact with the air. The advice generally given is to bury a considerable length of the rod in coke (not charcoal), and the plan is a very good one. Whether iron or copper is employed, it will be well to sprinkle the coke copiously with a strong solution of washing soda, for the purpose of neutralizing any acids that might corrode the metals. If a trench ten feet long be sunk to the depth of permanent moisture and filled to a depth of twelve inches with coke, it will be ready to receive the end of the rod, and will furnish a path for all the electricity that will ever tend to escape from the clouds to the earth.

As great mistakes have been often committed in this matter, we will point out a few of them. The general rule being to let the rod end in water, rods have been carried into cement cisterns, and cisterns hollowed by human means out of the solid rock. In all such cases the end of the rod was virtually insulated by the walls of the cistern, and such rods would be worthless. It is a common practice amongst lightning-rod men to form the earth termination by simply driving a crow-bar into the earth, and inserting the end of the rod in the hole thus formed. No reliance can be placed on an earth connection made in this manner. The crow-bar may have been driven into perfectly dry sand ; and, in any event, the amount of surface exposed by the rod, and the consequent section of earth brought into action, is altogether too small. Neither should rods be inserted into open wells of water, and especially should we avoid doing this in the case of wells that supply drinking-water. If the rod be made of copper, the well may be poisoned ; and, in any event, the combined action of the air and water cannot fail to corrode and injure the rod.

The accompanying engraving shows a new attachment which formed the subject of a recent commendatory report from the Committee on Science and Arts of the Franklin Institute, and is called—though precisely why, it puzzles us to tell—an Equilibrium Disk. The usual size of the disk is two feet in diameter. Its weight, including the seventy-two horizontal and perpendicular discharging-points, is about forty pounds. It is made of one of the common metals, though copper is preferred. Iron, however, being cheapest, is generally used. Regarding the mode of using it, it is stated that it should be sunk into the ground deep enough to be surrounded by perpetual moisture, rarely less than six feet. By exact adjustment, the rod passes through its centre, and is firmly solidified to the disk by copper surrounding. The fact that this device has been endorsed by a Committee of the Franklin Institute gives it a certain claim to public attention, and will doubtless cause it

FIG. 5.

to be extensively adopted; but the inventor of this attachment must have very confused ideas in regard to the action of electricity and its relation to points. That points serve to discharge electricity into the air by imparting it to the molecules of the air, which are then repelled and fly off, giving place to others, we can very well understand. But, when a point is embedded in a solid mass whose particles are immovable, we cannot see how a point can act more efficiently than any other surface. In so far as this new attachment presents a great extent of surface, it is to be commended; but we must remember that the same extent of surface might be obtained in a far more efficient form, and at a greatly diminished cost. Moreover, while a single point, exposed to the air, acts in the manner which the inventor of this device assumes that his points will act, a multiplicity of points will, as we have elsewhere stated, act pretty much like a ball. We refer to the device only for the purpose of condemning it.

Hints to Persons Exposed to Thunderstorms.

Little that is really of value can be said upon this subject. The old directions about avoiding the shelter of trees are sound ; though the advice which goes to an opposite extreme, and recommends us to go to the middle of an open field, is decidedly bad. A tall tree will, in general, protect a space of considerable diameter—the accepted rule giving a circle whose diameter is four times its height ; though, as elsewhere noted, we do not place implicit confidence in this. Now, it is a curious law of electrical action, that, if two bodies combine to form the medium of conduction, one being a good conductor and the other a poor one, the good conductor does not carry off *all* the charge ; it merely divides it with the poor conductor, which gets a share that is precisely equal to its capacity. When a man stands beside a tree that is struck by lightning, the charge divides, and the man is apt to receive such a proportion as is sufficient to kill him. But, if he had stood at a distance a little greater than the height of the tree, the latter would have attracted the bolt, which in that case would not have been divided. When standing in the middle of an open field, there is nothing to attract the lightning *from* us, and we run great risk of being struck.

The old directions about feather beds, glass windows, etc., are all nonsense. One of the safest of all places is a house well protected by good lightning-rods ; one of the most dangerous is a barn filled with new hay and without a rod. In a house unfurnished with lightning-rods, the most dangerous places are near the fireplace or chimney, or near those corners down which the water-spouts descend. Referring to our remarks about the division of a lightning-stroke when two parallel paths are presented to it, it must be noted that the quantity of the charge that passes down each conductor will be in proportion to its conducting power. A good lightning-rod has a power of conduction so much in excess of the body of a man, that there would be very little left to pass through him. On the other hand, the conducting power of trees, the soot of chimneys, tin water-pipes, bell-wires, etc., is so low, that a human body might have assigned to it a greater proportion of the charge than it could bear. A few years ago some wise-acre invented a portable lightning-rod consisting of an umbrella having a metal stem, to the lower part of which was attached a chain that was allowed to drag along the ground. Such a contrivance is not only useless, but dangerous. It would be utterly impossible to give such a portable rod a good ground connection ; and without this a lightning-rod is worse than useless.

The carrying of metal rods or bars is, of course, dangerous, because every mass of metal tends to open up a line of least resistance of which it will form a part ; though metal cannot be said to *attract* lightning in the sense that a magnet attracts a bar of iron.

It has been said that a lightning-rod no more attracts the lightning than a water-spout on a house attracts the rain. This is but partially true. The rod attracts the lightning to about the same extent that an open sluice may be said to attract the escaping waters of a pond.

Lightning-Rod Swindles.

—Next to the substitution of saw-dust packages for counterfeit money, and the sale of brass jewelry, the business of putting up lightning-rods is a favorite field for the operations of the swindling fraternity. Indeed, to such an extent is it carried, that at the West there are large companies formed whose gains are derived chiefly from the swindling part of their schemes, and not from the legitimate business of honestly putting up good rods. These scoundrels operate in a manner somewhat like this : Approaching a house in the absence of the men, they so terrify the ladies by means of horrible stories of persons killed and houses destroyed by lightning that when the proprietor returns he gets no peace until a rod is ordered. Then these fellows get an order for a rod at so much per foot, they giving a verbal estimate of the amount required, which estimate is not embodied in the contract, and forms no part thereof. They then go to work ; and being, by the terms of the contract, the sole judges of the amount required, they contrive to put on three or four times the number of feet originally mentioned. Of course, when the bill is presented, the victim sees no way of escape from the terms of his written contract, and he pays the amount ; and, when we remember that for a rod which costs them ten cents per foot, these fellows frequently charge 50 or 70 cents, it will be readily seen that the transaction is a profitable one for them. There is no doubt but that many of these suits might be successfully resisted, though of course the best way is to avoid falling into the trap ; and, to enable our readers to escape it, we offer the following hints :

If you make a bargain with a professional lightning-rod man to put up a rod for you, see that the contract specifies the *number of feet* as well as the price per foot. In your selection of the kind of rod, you will probably be guided by what we have previously said on this subject. If so, then remember that the best copper rod need not cost, for materials alone, over fifteen to twenty cents per foot ; and, if you have read attentively our paragraph on the arrangement of the rod, you will find no difficulty in determining where it should be carried, and, consequently, the number of feet that will be required.

The point in which you will be most apt to be cheated will be the ground connection. We have already given abundant directions in regard to this matter. Your only chance for safety lies in giving this part of the work your personal supervision during its progress ; after it is finished, no man can tell whether or not it is efficient, without such an examination as would cost as much as it would to do the work over again.

Surveying Lightning-Rods for Insurance Purposes.

—Insurance companies cannot, with safety, ignore the value

of the lightning-rod. We therefore give a few hints, which may prove of value to insurance agents who wish to determine whether a house or barn is efficiently protected.

The general principles governing the subject have been already discussed; but it must be borne in mind that a rod may be efficient, and yet neither economical nor neat. Moreover, while, for purposes of fire insurance, insulators are of no benefit, they are not a source of danger as they unquestionably are in regard to life. Points, etc., are not of much consequence if the rod has been properly arranged; in regard to this we would refer to the paragraph on that subject. So far as that portion of the rod that is above ground is concerned, the chief conditions to be observed are, that it shall be large enough, that it shall be continuous, and that it shall be well arranged. It is in regard to the ground connection that the great difficulty arises. To *examine* a ground connection would be a work of great labor and expense; and therefore we must rely upon other sources of information. We see no safe mode of obtaining the required information except by the written affidavit of the parties concerned. We lay great stress upon the importance of obtaining a *written affidavit*; mere verbal communications, and statements made in writing without the solemnity of an affidavit, are in general too vague to be depended upon.

The points of inquiry should be in regard to the depth to which the rod is carried, the mode in which it is arranged, and the extent of surface that it presents to the soil. After these points have been testified to, the surveyor should determine, by examining the wells in the vicinity, and by other indications, the probable depth at which permanent moisture is to be found; and from these data the efficiency of the ground connection can be determined. Any falsification in the statement made by the insured ought to void the policy; and, if an accident should occur from lightning, no time ought to be lost in making a thorough examination.

APPENDIX.

HOME-MADE LIGHTNING-RODS.

Since the issue of this little volume, we have received numerous inquiries in regard to the best material, etc., for the construction of Lightning-Rods, by those who do not make a regular business of putting them up. It frequently happens that farmers, builders, carpenters, and mechanics in general, have occasion to put up Lightning-Rods, and find a difficulty in procuring the necessary material and fixtures for the purpose. To such persons the following hints may prove useful.

In very out-of-the-way places, the rod may be made of round or square iron rod, five eighths of an inch in diameter. The lengths may be welded together in a blacksmith's shop, and drawn to the building for which it is intended. A single rod, leading by the shortest path from the highest chimney to the ground, is all that need be made in one piece. Those parts of the rod that lie along the ridge and the gable, may be connected to the main rod simply by a hook,—which should, however, be closed as tightly as possible upon the rod which it clasps; and if such joints occur over straw, shingles, or boards, it will be well to slip a piece of sheet-iron, tin, or zinc, under the joint, and secure it in its place by a few tacks. No fear need be entertained that the electricity will not follow the rod across such joints. The electricity produced by rubbing a common glass bottle with a piece of flannel will jump across a break half an inch wide; and it is therefore ridiculous to suppose that a break of a hundredth of an inch would impede the progress of a thunderbolt. But wherever such a break exists, a spark may be produced, and it is to prevent the possibility of such a spark setting fire to light combustibles that we recommend the use of thin sheet-metal laid under the joint.

But where good copper wire can be readily procured, it forms very much better material for the construction of home-made Lightning-Rods. The most suitable size is No. 0 or 00, which gives a solid section that is several times that of the tubular copper-rods in common use, most of which, by the way, are so light that they can not afford efficient protection against a heavy stroke. But notwithstanding the increased weight and efficiency of the plain wire, it will be found to cost much less *per foot* than the fancy rods usually sold by itinerant venders. Such wire can be obtained of any respectable hardware merchant, or it may be procured direct from the manufacturers, of whom there

are several,—amongst others the Waterbury Brass Company, whose
agency is at 52 Beekman Street, New York. The wire comes in rolls
measuring thirty to forty feet in length. The following gives the
diameters and weights per hundred feet of the sizes suitable for Lightning-Rods:—

Number on wire-guage.	Diameter.	Weight of 100 feet.
000	.409 inches.	50 lbs.
00	.364 "	40 "
0	.324 "	32 "
1	.289 "	25½ "
2	.257 "	20 "

Joints are easily made by flattening the ends of the pieces to be
united, binding them together with fine copper wire, and soldering the
whole, binding-wire and all, into a solid mass,--a very simple and easy
job. For attaching such a rod to a house, nothing is better than small
wire staples. A flat strip may be made to look much better than a
wire, but is not in any respect more efficient.

From the fact that it is much more pliable than copper wire, and
can be obtained in pieces of almost any length, the copper wire-rope
that is now so extensively used for many purposes, is an excellent
material for lightning-rods. It can be put up with less trouble than
any other kind of material, and it is the most convenient for forming
a ground connection; for, by untwisting the ends for a length of several feet, it is easy to spread the rope over a large surface of earth.

Rods carefully constructed of any of these materials will prove
far superior to those in common use, besides being far cheaper. Of
course all the precautions detailed in the body of this work in regard
to arrangement, ground connection, etc., must be fully carried out, if
we would secure perfect protection. In general, most of the details
are sufficiently well attended to, with the exception of the ground
connection. At this point, almost all the lightning-rods ordinarily
put up, fail. In our first edition we dwelt upon this point very fully,
and further observation not only fully confirms the remarks made on
page 23, but leads us to believe that our statements were not half
strong enough. Mr. Brooks, the well-known inventor of the Paraffin
Insulator, a man whose practical acquaintance with the science of
electricity is perhaps as extensive as that of any man in the country,
writes me as follows: "There were three disasters in this city (Phila-
delphia) last season, involving a loss of over two hundred and fifty
thousand dollars, by lightning—Solm's Woollen Factory, Morris Iron
Foundry, and the Point Breeze Refining Co., every one of which had
a lightning-rod; and the cases of general interest in the country are
of such common occurrence that I have, as a matter of interest, measured the resistance of these joints to earth, and find the average to be
above the resistance of a hundred miles of telegraph-wire." That is
to say, a flash of lightning would rather travel one hundred miles
along a telegraph-wire than pass from these lightning-rods to the
ground ! Can we wonder that buildings furnished with lightning-rods
are so often injured when struck?

Mr. Brooks advises that in cities the lightning-rods should be connected with the gas or water pipes. So far as the latter are concerned, there can not be much objection, but in regard to a connection with the gas-pipes, we hold different views. We must remember that there are two sides to this question, and two parties to be consulted,—the owner of the lightning-rod and the gas company. Looking wholly to the interest of the former, we would advise him by all means to connect his rod to the gas or water pipes, or both, but to be sure that the connection is made *outside* of his own premises, as instances have occurred where the pipes connected with the metre have been dislocated, and the gas set on fire by a lightning stroke. If, however, the rod be carefully connected with the pipes outside of the house, no danger can accrue to the house from this source. But how is it with the pipes themselves? Those who are familiar with the character of our streets, at the depth at which ordinary gas-pipes are laid, know that the soil is generally about as dry as soil can be, and it is kept in this condition by the pavement above, which shields it from the rain, and by the drains below, which carry off any little moisture that may find its way below the surface. These pipes, therefore, are embedded in an almost perfect non-conductor, and any electricity that enters them must flow along to a considerable distance before it can find an efficient outlet. In the mean time, it meets with an obstacle at every joint, for these joints are always made in such a way as to be virtually non-conductors. At each joint of the pipes, therefore, an electric explosion will occur; and, if the electricity be in sufficient quantity, the joint may be entirely ruptured, and instances are on record where gas-pipes have been disjointed, and the gas allowed to escape. While, therefore, a connection with the gas-mains must prove very advantageous to the individual whose house is to be protected, it may prove very disastrous to the public at large. We hold, therefore, that no man has a right to connect his lightning-rod with the gas-pipes of our cities. Let each one find a ground connection for himself, unless indeed two or more neighboring householders should decide to combine for the formation of a thoroughly efficient ground termination. Where four parties combine for this purpose, the advantages would be quadrupled, and, at the same time, there are fortunately no counterbalancing disadvantages.

In the instructions issued by the Inspector-General of Fortifications, British army, which instructions have been adopted in the Ordnance Manual published by the authority of the United States Government, and are substantially the same as those contained in our own little work, the following passage occurs: "If it be possible, in regulating the surface-drainage, to lead a flow of water, during the rain which generally accompanies thunder-storms, over the sites of the trenches in which the ground terminations are laid, it will be an additional precaution." The advice deserves attention.

OF

Books and Periodicals

PUBLISHED AND FOR SALE BY

THE INDUSTRIAL PUBLICATION COMPANY,

14 Dey Street, New York.

☞ *Any of these Books may be obtained from any Bookseller or Newsdealer, or will be sent Free by mail to any part of the United States or Canada ON RECEIPT OF PRICE.*

The Amateur's Handbook of Practical Information,

For the Workshop and the Laboratory. Second Edition.
Greatly Enlarged. Neatly Bound   -   - 15 cents.

This is a handy little book, containing just the information needed by Amateurs in the Workshop and Laboratory. Directions for making Alloys, Fusible Metals, Cements, Glues, etc.; and for Soldering, Brazing, Lacquering, Bronzing, Staining and Polishing Wood, Tempering Tools, Cutting and Working Glass, Varnishing, Silvering, Gilding, Preparing Skins, etc., etc.

The New Edition contains extended directions for preparing Polishing Powders, Freezing Mixtures, Colored Lights for tableaux, Solutions for rendering ladies' dresses incombustible, etc. There has also been added a very large number of new and valuable receipts.

Rhymes of Science: Wise and Otherwise.

By O. W. Holmes, Bret Hart, Ingoldsby, Prof. Forbes, Prof. J. W. McQ. Rankine, Hon. R. W. Raymond, and others.

With Illustrations. Cloth, Gilt Title.   -   50 cents

Section Cutting.

A Practical Guide to the Preparation and Mounting of Sections for the Microscope; Special Prominence being given to the Subject of Animal Sections. By Sylvester Marsh. Reprinted from the London edition. With Illustrations. 12mo., Cloth, Gilt Title. - 75 cents.

This is undoubtedly the most thorough treatise extant upon section cutting in all its details. The American edition has been greatly enlarged by valuable explanatory notes, and also by extended directions, illustrated with engravings, for selecting and sharpening knives and razors.

A Book for Beginners with the Microscope.

Being an abridgment of "Practical Hints on the Selection and Use of the Microscope." By John Phin. Fully illustrated, and neatly and strongly bound in boards. 30 cts.

This book was prepared for the use of those who, having no knowledge of the use of the microscope, or, indeed, of any scientific apparatus, desire simple and practical instruction in the best methods of managing the instrument and preparing objects.

How to Use the Microscope.

A Simple and Practical Book, intended for beginners. By John Phin, editor of "The American Journal of Microscopy." Second Edition. Greatly Enlarged, with 50 illustrations in the text and 4 full-page engravings printed on heavy tint paper. 12mo., Neatly bound in Cloth, Gilt Title. - - - - 75 cents.

The Microscope.

By Andrew Ross. Fully Illustrated. 12mo., Cloth, Gilt Title. - - - - - 75 cents.

This is the celebrated article contributed by Andrew Ross to the "Penny Cyclopædia," and quoted so frequently by writers on the Microscope. Carpenter and Hogg, in the last editions of their works on the Microscope, and Brooke, in his treatise on Natural Philosophy, all refer to this article as the best source for full and clear information in regard to the principles upon which the modern achromatic Microscope is constructed. It should be in the library of every person to whom the Microscope is more than a toy. It is written in simple language, free from abstruse technicalities.

Diatoms.

' Practical Directions for Collecting, Preserving, Transporting, Preparing and Mounting Diatoms. By Prof. A. Mead Edwards, M. D., Prof. Christopher Johnston, M. D., Prof. Hamilton L. Smith, LL. D.

12mo., Cloth. - - - - 75 cents.

This volume undoubtedly contains the most complete series of directions for Collecting, Preparing and Mounting Diatoms ever published. The directions given are the latest and best.

Common Objects for the Microscope.

By Rev. J. G. Wood. Upwards of four hundred illustrations, including twelve colored plates by Tuffen West.

Illuminated Covers. - - - 50 cents.

This book contains a very complete description of the objects ordinarily met with, and as the plates are very good, and almost every object is figured, it is a most valuable assistant to the young microscopist.

Five Hundred and Seven Mechanical Movements.

Embracing all those which are Most Important in Dynamics, Hydraulics, Hydrostatics, Pneumatics, Steam Engines, Mill and Other Gearing, Presses, Horology and Miscellaneous Machinery; and including Many Movements never before published, and several of which have only recently come into use. By Henry T. Brown, editor of the "American Artisan." Eleventh Edition. $1.00.

This work is a perfect Cyclopædia of Mechanical Inventions, which are here reduced to first principles, and classified so as to be readily available. Every mechanic that hopes to be more workman, ought to have a copy.

The Six Days of Creation.

The Chemical History of the Six Days of Creation. By John Phin, C. E., editor of "The American Journal of Microscopy." 12mo., Cloth. - - 75 cents.

Stories About Horses.

Just the Book for Boys. With eight full-page engravings. In Boards, 25 cents. In Cloth - 50 cents.

Shooting on the Wing.

Plain Directions for Acquiring the Art of Shooting on the Wing. With Useful Hints concerning all that relates to Guns and Shooting, and particularly in regard to the art of Loading so as to Kill. To which has been added several Valuable and hitherto Secret Recipes, of Great Practical Importance to the Sportsman. By an Old Gamekeeper.

12mo., Cloth, Gilt Title. - - - 75 cents.

The Pistol as a Weapon of Defence,

In the House and on the Road.

12mo., Cloth. - - - - - 50 cents.

This work aims to instruct the peaceable and law-abiding citizens in the best means of protecting themselves from the attacks of the brutal and the lawless, and is the only practical book published on this subject. Its contents are as follows: The Pistol as a Weapon of Defence.—The Carrying of Fire-Arms.—Different kinds of Pistols in Market; How to Choose a Pistol.—Ammunition, different kinds; Powder, Caps, Bullets, Copper Cartridges, etc.—Best form of Bullet.— How to Load.—Best Charge for Pistols.—How to regulate the Charge.—Care of the Pistol; how to Clean it.—How to Handle and Carry the Pistol.—How to Learn to Shoot.—Practical use of the Pistol; how to Protect yourself and how to Disable your antagonist.

Lightning Rods.

Plain Directions for the Construction and Erection of Lightning Rods. By John Phin, C. E., editor of "The Young Scientist," author of "Chemical History of the Six Days of the Creation," etc. Second Edition. Enlarged and Fully Illustrated.

12mo., Cloth, Gilt Title. - - - 50 cents.

This is a simple and practical little work, intended to convey just such information as will enable every property owner to decide whether or not his buildings are thoroughly protected. It is not written in the interest of any patent or particular article of manufacture, and by following its directions, any ordinarily skilful mechanic can put up a rod that will afford perfect protection, and that will not infringe any patent. Every owner of a house or barn ough' to procure a copy.

Instruction in the Art of Wood Engraving.

A Manual of Instruction in the Art of Wood Engraving; with a Description of the Necessary Tools and Apparatus, and Concise Directions for their Use; Explanation of the Terms Used, and the Methods Employed for Producing the Various Classes of Wood Engravings. By S. E. Fuller.

Fully illustrated with Engravings by the author, separate sheets of engravings for transfer and practice being added.

New Edition, Neatly Bound.   -   -   30 cents.

What to Do in Case of Accident.

What to Do and How to Do It in Case of Accident. A Book for Everybody. 12mo., Cloth, Gilt Title. 50 cents.

This is one of the most useful books ever published. It tells exactly what to do in case of accidents, such as Severe Cuts, Sprains, Dislocations, Broken Bones, Burns with Fire, Scalds, Burns with Corrosive Chemicals, Sunstroke, Suffocation by Foul Air, Hanging, Drowning, Frost-Bite, Fainting, Stings, Bites, Starvation, Lightning, Poisons, Accidents from Machinery, and from the Falling of Scaffolding, Gunshot Wounds, etc., etc. It ought to be in every house, for young and old are liable to accident, and the directions given in this book might be the means of saving many a valuable life.

BOUND VOLUMES OF

The Technologist, or Industrial Monthly.

The eight volumes of THE TECHNOLOGIST, OR INDUSTRIAL MONTHLY, which have been issued, form a Mechanical and Architectural Encyclopædia of great value; and, when properly bound, they form a most important addition to any library. The splendid full-page engravings, printed on tinted paper, in the highest style of the art, are universally conceded to be the finest architectural and mechanical engravings ever published in this country. We have on hand a few complete sets, which we offer for $16.00, handsomely and uniformly bound in cloth.

We have also a few extra sets of Vols. III to VIII inclusive. These six volumes we offer for $8.00 bound in cloth. As there are but a very few sets remaining, those who desire to secure them should order immediately.

NOTE.—The above prices do not include postage or express charges. The set weighs altogether too much to be sent by mail.

Just Published. 1 Vol., 12mo. Neatly Bound in Cloth, Gilt Title. Price 75 cents.

HOW TO USE THE MICROSCOPE.

A Simple and Practical Book, intended for beginners.

By JOHN PHIN,

Editor of " The American Journal of Microscopy."

Second Edition. Greatly Enlarged, with 50 illustrations in the text, and 4 full-page engravings printed on heavy tint paper.

CONTENTS:

WHAT A MICROSCOPE IS.—Different Kinds of Microscopes.—Simple Microscopes.—Hand Magnifiers.—The Coddington Lens.—The Stanhope Lens.—Raspail's Microscope.—The Excelsior Microscope.—Twenty-five cent Microscopes and how to make them.—Penny Microscopes.

COMPOUND MICROSCOPES.—Different kinds of Objectives.—Non-Achromatic Objectives.—French Achromatic Objectives—Objectives of the English Form.—Immersion Objectives—Focal Lengths corresponding to the numbers employed by Nachet, Hartnack and Gundlach.

HOW TO CHOOSE A MICROSCOPE.—Microscopes for Special Purposes.—Magnifying Power required for different purposes.—How to judge of the quality of the different parts of the Microscope.

ACCESSORY APPARATUS.—Stage Forceps, Animalcule Cage, etc.

ILLUMINATION.—Sun Light.—Artificial Light.—Bulls-Eye Condenser.—Side Reflector.—The Lieberkuhn.—Axial Light.—Oblique Light.—Direct Light.

HOW TO USE THE MICROSCOPE.—How to Care for the Microscope.

HOW TO COLLECT OBJECTS.—Where to find Objects.—What to Look for.—How to Capture Them.—Nets.—Bottle-Holders.—Spoons.—New Form of Collecting Bottle.—Aquaria for Microscopic Objects.—Dipping Tubes.

THE PREPARATION AND EXAMINATION OF OBJECTS.—Cutting Thin Sections of Soft Substances.—Sections of Wood and Bone.—Improved Section Cutter.—Sections of Rock.—Knives.—Scissors.—Needles.—Dissecting Pans and Dishes.—Dissecting Microscopes.—Separation of Deposits from Liquids.—Preparing whole Insects.—Feet, Eyes, Tongues, Wings, etc., of Insects.—Use of Chemical Tests.—Liquids for Moistening Objects.—Refractive Power of Liquids.—Covers for Keeping out Dust.—Errors in Microscopical Observations.

PRESERVATION OF OBJECTS.—General Principles.—Recipes for Preservative Fluids.—General Rules for Applying them.

MOUNTING OBJECTS.—Apparatus and Materials for: Slides, Covers, Cells, Turn-Table, Cards for Making Cells, Hot-Plate, Lamps, Retort Stand, Slide-Holder, Mounting Needles, Cover Forceps, Simple Form of Spring Clip, Centering Cards, Gold Size, Black Japan, Brunswick Black, Shellac, Bell's Cement, Sealing Wax Varnish, Colored Shellac, Damar Cement, Marine Glue, Liquid Glue, Dextrine.—Mounting Transparent Objects Dry.—Mounting in Balsam.—Mounting in Liquids.—Mounting of Whole Insects.—How to Get Rid of Air-Bubbles.—Mounting Opaque Objects.

FINISHING THE SLIDES.

May be obtained from any Bookseller or News Agent, or will be sent by mail, postage paid, on receipt of price.

THE AMERICAN
JOURNAL OF MICROSCOPY,
AND
POPULAR SCIENCE.

PROSPECTUS.

The object of the JOURNAL OF MICROSCOPY is to diffuse a knowledge of the best methods of using the Microscope; of all valuable improvements in the instrument and its accessories; of all new methods of microscopical investigation, and of the most recent results of microscopical research. The JOURNAL does not address itself to those who have long pursued certain special lines of research, and whose wants can be supplied only by elaborate papers, which, from their thoroughness, are entitled to be called monographs rather than mere articles. It is intended rather to meet the wants of those who use the microscope for purposes of general study, medical work, class instruction, and even amusement, and who desire, in addition to the information afforded by text-books, such a knowledge of what others are doing as can be derived only from a periodical. With this object in view, therefore, the publishers propose to make the JOURNAL so simple, practical and trustworthy, that it will prove to the advantage of every one who uses the microscope at all to take it.

ILLUSTRATIONS —The JOURNAL will be freely illustrated by engravings representing either objects of natural history or apparatus connected with the microscope.

TRANSACTIONS OF SOCIETIES.—THE AMERICAN JOURNAL OF MICROSCOPY is not the organ of any Society, but it gives the proceedings of all Societies whose officers send us a report. As the JOURNAL is devoted *wholly* to Microscopy, and is in good form and size for binding, no better medium can be had for preserving the scientific records of any society. Matters of mere business routine we are frequently obliged to omit for want of room.

EXCHANGES.—An important feature of the JOURNAL is the exchange column, by means of which workers in different parts of the country are enabled, without expense, except for postage, to exchange slides and materials with each other.

TERMS.

During the first two years of its existence, the subscription to the AMERICAN JOURNAL OF MICROSCOPY was only fifty cents per year, but at the request of more than two-thirds of the subscribers, the size of the JOURNAL has been doubled, and the price raised to

ONE DOLLAR PER YEAR.

Four copies for three dollars. Those who wish to economize in the direction of periodicals, would do well to examine our clubbing list.

FOREIGN SUBSCRIBERS.—The JOURNAL will be sent, postage paid, to any country in the Postal Union for $1.24, or 5 shillings sterling per year. English postage stamps, American currency or American postage stamps taken in payment. In return for a postal order or draft for £1 5s., five copies of the JOURNAL will be furnished and mailed to different addresses. Make all drafts and postal orders payable to John Phin.

BACK VOLUMES.—We have on hand a few copies of Vols. I and II, bound in handsome cloth cases, which we offer for $1 25 each. Vols. I and II, bound, and the numbers of Vol. III, as issued, we offer for $2.50. We can no longer supply complete sets of 1876-7 in sheets. To those who wish to examine the journal, we will send ten odd numbers for 25 cents.

Advertisements.

The JOURNAL OF MICROSCOPY, from its very nature, is a visitor to the very best families, and its value as an advertising medium has therefore proved to be much above that of average periodicals. A few select advertisements will be inserted at the rate of 30 cents per line, nonpariel measure, of which twelve lines make an inch. Address

AMERICAN JOURNAL OF MICROSCOPY,
P. O. Box 2852, New York.

NEW BOOKS IN PRESS.

Supplements to this Catalogue will be issued from time to time, and will be sent free to any one desiring them. Since our removal to the larger and more commodious premises at 14 Dey Street, we have made arrangements to greatly extend our publishing business.

The following books are in a forward state of preparation, and will be issued at an early day:

The Aquarium.

A Practical Treatise. With nearly one hundred illustrations. By. A. W. Roberts, formerly of Barnum's Museum Aquaria, Collector for the New York Aquarium, and Superintendent of Sea Side Aquarium.

Mr. Roberts having had over thirty years experience, has therefore been able to produce a thoroughly practical book. The accuracy of the illustrations are guaranteed, by the fact that they are carefully drawn and engraved by the author from life.

Taxidermy; or, The Art of Preserving and Stuffing Beasts, Birds and Fishes.

This work is by a practical Taxidermist, who is perhaps better known than any other of the profession in this country. It will be simple, thorough and practical.

Fret and Scroll Saws: Their Construction and Use.

Notwithstanding the many works on this subject now before the public, we think that there is room for one more. The volume which we propose to bring out is by a gentleman whose profession (that of an architect) has given him special qualifications for producing a really good book, and one free from the errors of taste and instruction in practical workmanship, which have disfigured so many attempts in this direction. It will be very copiously illustrated.

We shall also issue a portfolio of twelve designs, full working size, for Fret and Scroll Saw Work. The set will be sold at a very moderate price.

Young Scientist Companion.

This is a small pocket book intended for the practical man, as well as the young scientist. It will contain rules and data for calculations; tables of physical and chemical data; recipes; processes, and in fact just that practical information which every man has so often occasion to refer to.

Cage Birds, and How to Care for Them.

With a full account of the best methods of training, etc.

This work is by a lady who has long been known as a successful authoress, and who is noted for her skill in the care of household pets. It will be splendidly illustrated with nearly one hundred engravings, and cannot fail to find its way into every home where a canary or other song bird is kept.

Manual of Urinary Examination, Chemical and Microscopical.

For the Use of Physicians, Medical Students and Clinical Assistants. By Frank M. Deems, M. D., Laboratory Instructor in the University Medical College, New York; Member of the Medical Society of the County of New York; Member of the New York Microscopical Society etc., etc.

How to Learn to Draw.

By an old Teacher of Drawing.

This is the first of a series of small practical books to be called "The Young Scientist Manuals." It does not profess to teach the science or art of Drawing; it is intended rather to point out to the beginner the best methods of going to work. It is intended for self-taught students, and contains a great deal that is not to be found in the books, but is generally imparted to the pupil directly by the teacher.

The price of this series has been fixed at 20 cents each, neatly and strongly bound in boards.

CATALOGUE OF MICROSCOPES.—We are preparing a CATALOGUE OF MICROSCOPES, with hints in regard to the selection of a microscope, especially as regards its suitability for the various purposes for which microscopes are used. A copy of this catalogue will be sent free to any one applying for it. Applications received before the Catalogue is published will be carefully registered, and filled as soon as possible.

NEW WORKING MICROSCOPE.

Manufactured by Geo. Wale.

Price—With two Eye-Pieces, 2-3 and 1-5 Objectives, Iris Diaphragm, and Black Walnut Case, - - - $35.

INDUSTRIAL PUBLICATION COMPANY,
14 Dey Street, New York, Sole Agents.

THE NEW WORKING MICROSCOPE.

This Microscope has just been brought out by Mr. Geo. Wale, whose reputation as a maker of fine stands is so well known. It embodies several new and important features, foremost amongt which is the method of hanging the body, so that it may be made to incline at any angle. The method now in general use for this purpose changes the position of the centre of gravity of the instrument, and renders the microscope more or less unsteady. The new method avoids this difficulty, and at the same time furnishes a secure and convenient means of clamping the body at any position—a point of considerable importance, when the instrument is used for some purposes.

The stage is of a new construction, very thin, so as to admit the greatest obliquity in the illumination of objects, and with clips which move round it, thus giving many of the advantages of a rotary stage. The clips may be easily and quickly removed, so as to leave a clear stage, and they may also be so applied as to hold the slide against the *under* side of the stage, when very oblique light is required for resolving difficult test objects.

There are two means of adjusting for focus, a coarse movement by means of a well-made rack and pinion, and a fine movement, in which the entire body is moved by what is generally called a "micrometer" screw, acting on a lever. The latter movement has this great advantage, that it does not change the distance between the eye-piece and the objective (as is the case with most of the English fine movements), and consequently does not vary the magnifying power of the instrument—an important point in making delicate micrometric measurements.

The diaphragm is of the Iris pattern, a form which is generally acknowledged to be the best, but which has hitherto been very costly, those usually supplied being sold for $16. This Iris diaphragm is a new form, which, with several other features of this stand, has been patented by Mr. Wale. It may be easily and quickly applied or removed.

The New Working Microscope is of substantial make, elegant design, and thoroughly good workmanship. It has the Society screw; draw-tube, with Society screw at lower end for receiving long focus objectives, analyzing prism, etc.; plane and concave mirrors, the distance of which from the object may be varied; two Eye-pieces; 2-3 and 1-5 Objectives of Wale's Histological Series. The 1-5 easily and clearly resolves the *P. Angulatum* with light of a very slight degree of obliquity.

This Microscope, with draw-tube pulled out, and the body inclined, as shown in the figure, stands 13½ inches high; the draw-tube pulls out so as to give the standard length of body, 10 inches. When placed in a vertical position, with the draw-tube pushed in, the instrument stands only 10½ inches high, so that it can be used comfortably on an ordinary table. This is a great convenience in the rapid examination of liquids. It is very complete in all its appointments, is capable of receiving and doing justice to any accessories, and is contained in a handsome black walnut box, with brass handle, lock and key. It is equal to all the ordinary requirements of any physician, student or naturalist.

PRICE—With Accessories, as above described - $35.00.

THE YOUNG SCIENTIST,

A Practical Journal for Amateurs.

ISSUED MONTHLY.   **Price 50 Cents per year.**

It is characteristic of young Americans that they want to be DOING something. They are not content with merely *knowing* how things are done, or even with *seeing* them done; they want to do them themselves. In other words, they want to experiment. Hence the wonderful demand that has sprung up for small tool chests, turning lathes, scroll saws, wood carving tools, telegraphs, model steam engines, microscopes and all kinds of apparatus. In nine cases out of ten, however, the young workman finds it difficult to learn how to use his tools or apparatus after he has got them. It is true that we have a large number of very excellent text-books, but these are not just the thing. What is wanted is a living teacher. Where a living teacher cannot be found, the next best thing is a live journal, and this we propose to furnish. And in attempting this it is not our intention to confine ourselves to mere practical directions. In these days of knowledge and scientific culture, the "Why" becomes as necessary as the "How." The object of the YOUNG SCIENTIST is to give clear and easily followed directions for performing chemical, mechanical and other operations, as well as simple and accurate explanations of the principles involved in the various mechanical and chemical processes which we shall undertake to describe.

The scope and character of the journal will be better understood from an inspection of a few numbers, or from the list of contents found on a subsequent page, than from any labored description. There are, however, three features to which we would call special attention:

CORRESPONDENCE.—In this department we intend to place our readers in communication with each other, and in this way we hope to secure for every one just such aid as may be required for any special work on hand.

EXCHANGES.—An exchange column, like that which has been such a marked success in the *Journal of Microscopy*, will be opened in the YOUNG SCIENTIST. Yearly subscribers who may wish to *exchange* tools, apparatus, books, or the products of their skill, can state what they have to offer and what they want, *without charge.* Buying and selling must, of course, be carried on in the advertising columns.

ILLUSTRATIONS.—The journal will make no claims to the character of a "picture book," but wherever engravings are needed to make the descriptions clear they will be furnished. Some of the engravings which have already appeared in our pages are as fine as anything to be found in the most expensive journals.

Special Notice.

As our journal is too small and too low-priced to claim the attention of news dealers, we are compelled to rely almost wholly upon subscriptions sent directly to this office. As many persons would no doubt like to examine a few numbers before becoming regular subscribers, we will send four current numbers as a trial trip for

FIFTEEN CENTS.

CLUBS.

Where three or more subscribe together for the journal, we offer the following liberal terms:

3 copies for	....	$1.25
5 " "	....	2.00
7 " "	....	2.75
10 " "	....	3.50

Advertisements, 30 cents per line.

As postal currency has nearly disappeared from circulation, we receive postage stamps of the lower denominations (ones, twos and threes) at their full value. Postal orders are, however, much safer and more convenient. To avoid delay and mistakes address all communications to "THE YOUNG SCIENTIST, Box 2832, New York," and make all checks and orders payable to John Phin.

www.ingramcontent.com/pod-product-compliance
Lightning Source LLC
Chambersburg PA
CBHW032141270626
47172CB00009B/844

* 9 7 8 3 3 3 7 2 1 4 7 3 9 *